Humpback whale

Humpback whales

Whales swim

deep under water.

Whales jump

out of the water too!

Humpback whale

Whales swish

their big tails up
and down.

Most whales live

in groups.

Humpback whales feeding

Atlantic spotted dolphins

Dolphins swim

in the water.

Dolphins jump

into the air just
for fun!

Bottlenose dolphins

Bottlenose dolphin

Dolphins make

many different sounds.

Dolphin mothers

take good care
of their babies.

Bottlenose dolphins

Bottlenose dolphins

Dolphins like

to be together.

Find the Hidden Whales & Dolphins

The whales and dolphins you see here can be found repeated in earlier pages of this Ranger Rick Zoobies. Can you match up these ocean mammals 1-2-3-4-5 with their twins hiding elsewhere in Ranger Rick Zoobies? Plus, use the whale illustrations throughout this book to help your child learn to count!

Did you know that dolphins are actually whales? They are one of the several types of toothed whales. Whales that do not have teeth are called baleen whales.